The Dead Blue & You

By William (Bill) C. McElroy

ISBN 9781692123383

Preface:

Have you been to an ocean beach, either in the US or on foreign soil? Have you looked out at the curve of the earth, felt the warm sand under your feet, and listened to the waves gently breaking on the shore? If you have you know it is a special experience, and one that billions of humans may never experience.

What if the warm sand was covered with garbage, green or red algae, toxic chemicals, nuclear waste, human waste, dead creatures, tar and crude oil, and tons of plastic items; would you still enjoy the experience?

Our oceans are the lifeblood of our planet, and we are slowly killing each with our disregard for nature and the delicate balance that keeps our planet habitable. This short booklet is the author's attempt to have you, my reader, informed about our fragile condition and what may happen if we chose to not be informed and thus, not do something to protect ourselves.

Table of Contents:

Section - # 01 – Author:

This is a little background about the author and why he took the time to do the research, writing, and publishing of this book.

Farm Life:

I was raised on a farm in the Catskill Mountains of New York State in the 1940/50s. My parents had no knowledge of how to run a dairy farm, how to contour the land, grow food for livestock, or for us. But, they learned, and we had a few good years until Borden's changed the rules of milk farming and we switched to Potato farming. We also had chickens, pigs, cats, a dog, and a horse or two.

My father had been in the US Army and left to start the farm, but WWII and then the Korean War changed things and he left the farm operations to Mother and her very young son, me. With the exception of spreading manure on the garden and the fields, we were pretty much 'organic' farmers as the garden and the fields were not chemically fertilized or sprayed for insects. We did lose the entire potato crop, but it was due to poor soil and lack of knowledge, not insects or other animals.

I mention this because our farm did little damage to the environment and the food produced in our garden was fresh and delicious. The chicken and pork produced was juicy, flavorful, and even the fat was edible. Milk was creamy and delicious. We had tons of fresh fruit and nuts that became our snacks, and we canned everything we could; we even make our own jellies, jams, and candies. We did sell the farm when dad decided to stay in the US Army as a Nike Missile Site Repairman.

Robert D. Conrad Oceanographer:

After my four years in the US Air Force (AF30153A) I joined the Robert D. Conrad an oceanographic ship run by Lamont Geophysical Observatory a division of Columbia University and contracted by the US Navy. We circumnavigated the globe and crossed the equator and the International Date Line many times.

Our research was to sample the salinity of the oceans, take bottom soundings, study the earth's gravity fields, take core samples, study the plankton layers, test the first GPS system, find mineral deposits, look for submarine hiding spots, and study the earth's magnetic fields (My primary job). This scientific cruise took the good part of a full year, and was a great experience and adventure. {See my Amazon Kindle book "Pioneers of Oceanography: Saga of the Robert D. Conrad" by William (Bill) C. McElroy (Author)- Kindle Edition

It brings us to the next item, the Indian Ocean.

Indian Ocean River:

The day was mild and the sky a deep blue, very few clouds and no sign of storms. We were heading toward Colombo, Ceylon (Now

Sri Lanka) and were nearly 1,200 miles out from any land when we spotted what looked like a wave several thousand yards ahead.

As we got closer we saw floating lumber, logs, furniture, and trash of all sorts. We had to cross this wave of trash and when we did we saw the second wave about 800 feet in front of us; it too was a wave of various trash.

We stopped between the two waves and looked at the trash and the dozens of sea snakes, sharks, and other marine life that was feeding on the garbage that was floating in the waves. Taking water samples we determined that the water was not salt water, but freshwater and it was about 200 feet deep. We had found the Meghna River that came from Bangladesh and the trash that was from thousands of miles of riverfront as each river of the land passed through the local countryside.

The year was 1965 and this was a positive sign that man was polluting our oceans; and needed to take better care of his or her environment.

Section - # 02 – Our Oceans:

Our Blue planet is covered by water, lots of water that provides life to every living thing on the planet. Here is a little about it.

Our Seas & Oceans:

It is disputed, but the recent consensus is that there are five oceans on our planet, the Atlantic, Pacific, Indian, Arctic, and Antarctic. We also have several 'Seas' that included the Red Sea, the Black Sea, the Caspian, the Adriatic, the Mediterranean, and the Persian Gulf. Each of these are surrounded by land that is occupied by humans, and each is now facing chemical and trash destruction.

The Ocean Above:

Look up on most days and you will see clouds floating effortlessly by on their way to somewhere. These clouds are an ocean of moisture that usually originate from the bodies of water that make up about 70% of the earth. Our weather is a combination of air currents and moisture that create fair weather days, rainy days, and destructive stormy days. As the oceans and the lands heat or cool the amount of evaporation changes, and as the water evaporates it moves vertically from the surfaces, thus creating a void that sucks in air, and thus creating air movements or winds.

REF: https://www.thoughtco.com/what-is-the-hydrologic-cycle-1435330

Importance of Our Oceans:

Our oceans move from the equator to the poles in what is known as *Thermohaline Circulation* flow. As the water gets colder it gets denser and sinks, this pulls the warmer top water toward the void, and this motion along with the tilt of the earth provides us with the seasons and seasonal weather. As the cold water flows toward the equator it warms up and rises toward the surface, thus pulling more of the bottom cooler water toward it. This combination of warm and cold-water motions form a current loop that keeps the planet's weather reasonably stable.

Food Supply from Our Oceans:

Americans love their meat, beef, pork, chicken, lamb, turkey, and the variations of each. Much of the rest of the world depend on the oceans to provide their food, i.e. fish, shellfish, seaweed, and sea going mammals such as seals, whales, walrus, etc. We also get much of our salt from the oceans, as well as some minerals.

As the oceans warm and cool the creatures that live within move with the temperatures. Many varieties of marine creatures have over the centuries adapted themselves to certain temperatures, salinities, food sources, and such and therefore, when any of these factors change the creatures either have to move or perish; many perish as they just cannot adapt as fast as the changes occur.

How Our Oceans Affects Our Weather:

In addition to the vertical up and down flow of ocean waters, we also have the current loops that sweep past our coast. For example, the Atlantic current flows from the equator up the coast of Central America and Mexico to the coast of the United States and then across the southern border of Greenland, to the United Kingdom, down the coast of France, Portugal, and Spain, to western Africa, and then across the ocean back to South America and then upwards back to the United States east coast. This flow provides us with our summer and winter beach weather, and our potentials for gathering seafood and growing land based crops.

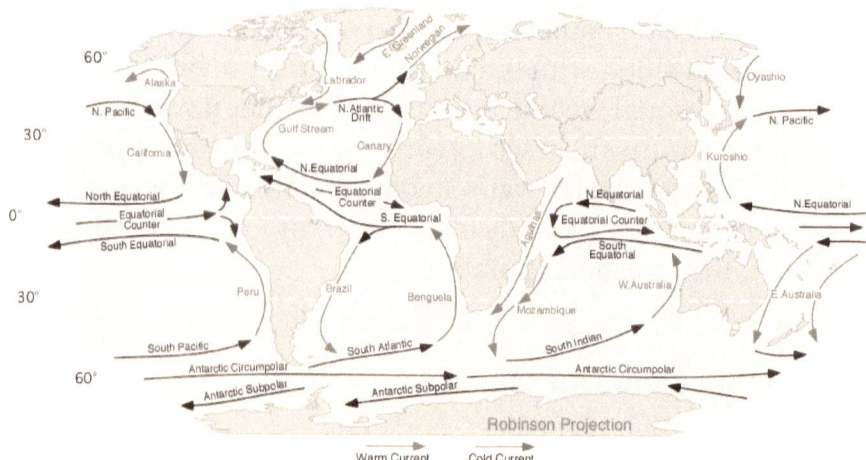

{Picture is Public Domain; it is from US Government Publications}

Climate Change and Our Weather:

Weather is transit and it can be sunny one second and raining like crazy the next. Many areas of our planet have weather that is created specifically to an area. As air and water vapor enter valleys, or climb mountains, each combines to from a localized weather pattern that is repeatable over a period of time.

Climate Change on the other hand is a long-term effect that creates changes in the weather patterns due to long term heating or cooling of a wide area of our planet. We are currently experiencing Climate Change due to the warming of our oceans and land areas from a 'baseline' period. NOAA (National Oceanic and Atmospheric Administration) uses a 1900 to 2000 period as the baseline, and considers that a long-term climate change can resulting a long-term change to a weather pattern.

REF: https://www.epa.gov/climate-indicators/climate-change-indicators-us-and-global-temperature

Section - # 03 – Pollution:

Pollution can be from chemicals, animal and human waste, storm destruction of man-made items, discarded man-made trash, and natural sources.

Greenhouse Effect and An Ice Age:

"In 1978 these same so called scientist SWORE we were headed into the next ICE AGE. NOW we are setting the planet on fire. "

This is a statement made by a person that has no scientific knowledge. First let's ask, why is Florida spending $400,000,000 to keep the rising sea from drowning its cities?

Now. --- As we put pollution into our air, we form a Greenhouse Effect that keeps heat in, and thus heats our waters, which in turn forms water vapor that keeps more heat in. As this slowly builds up, we will get hotter and hotter, but this extra heat will produce a thicker barrier to the sunlight that heats us, and eventually will block out the sun's heating rays, which will cause the earth to cool, and thus form an Ice Age. So, the scientists were and are 100% correct, Global Warming can produce both a heat wave and an Ice Age. It is all in the timing, and we are currently in the 'Heating Up' period, which will eventually result in the 'Cooling Down' period as our pollution blacks out the sun.

Farming Pollution:

Farmers are under pressure to produce crops or animals that can feed millions of people in the US, and in other nations. We have 6.3 billion people to feed, and many times that number of animals to feed. Thus the faster a farmer can bring a crop of food to market the better, and more profitable. To increase the speed of production the farmers are using more and more chemicals on their lands and in the feed that they provide the animals. These chemicals can be for preventing insect destruction of the crops, for forcing growth (Chicken), for producing fatter and heavier animals (Cattle), for ripening certain foods (Bananas), etc.

Manure has been used for centuries as a 'fertilizer' for both animal (US and many other nations) and human edible plant growth (China and Japan) and for the most part has been harmless to animals and humans. But, manure that is laced with dangerous chemicals (Growth Hormones or Insecticides) is harmful and we are beginning to experience the effects as more and more food borne diseases are detected.

Some Common Foodborne Germs:

The top five germs that cause illnesses from food eaten in the United States are: Norovirus, Salmonella, Clostridium perfringens, Campylobacter, and Staphylococcus aureus (Staph).

Many different disease-causing germs can contaminate foods, so there are many different foodborne infections.

CDC estimates that each year 48 million people get sick from a foodborne illness, 128,000 are hospitalized, and 3,000 die.

Researchers have identified more than 250 foodborne diseases.

Poor handling of our food crops causes some of these diseases, and some are due to the contamination by chemicals. Laws have been passed to help prevent these health problems, but many nations do NOT have sufficient laws or inspections, including the US.

Energy Production Pollution:

The two biggest producers of energy are the coal-fired electric power plants and the use of the combustion engine in our vehicles. We produce the energy by burning hydrocarbons, i.e. coal and refined oil. Each of these produce toxic fumes and toxic waste that gets into our water and air that we drink and breathe.

Back about the 13th century someone was cooking on an outcropping of black rock and found that the ground caught on fire. Later in India the same results were found. In the US coal was found along the Illinois River in 1679 by some French explorers. Mining in the US started in 1748 near Richmond, Virginia.

Coal is abundant in our soil where several millions of years ago there were swamps, forests, and lots of vegetation. When the vegetation died it was covered with earth from blowing dirt, volcanoes, meteor strikes, etc., and under intense pressure the carbon content solidified into what we know as coal. Coal is mostly Carbon, but may also contain hydrogen, oxygen, sulfur, and nitrogen in trace amounts. As it burns it produces fly ash, bottom ash, and flue-gas desulphurization sludge that may contain mercury, uranium, thorium, arsenic, and other heavy metals. It also produces carbon dioxide, carbon monoxide, and Sulphur Dioxide and in some instances Methane. Methane is also released from the earth when coal is mined.

"Dems are truly sick individuals. They devastate the Coal industry and then act like they care."

Ignorance of FACT does not mean that statements like this are true. The fact is that Coal Mining has changed from digging a hole with a shovel to leveling entire mountaintops. The industry today has less than 60,000 workers whereas; when they were digging holes it had many times that number of workers.

If you do not believe in progress, we would still be using horse and buggies for our primary transportation. We have advanced to Green Energy, i.e. Solar, Bio, Geo, Wind, etc., and the Coal industry is going to disappear. The people in the industry got sick on coal dust, killed by coal dust fires, polluted their water with coal dust, buried their towns in coal ash, etc., it is time for them to WAKE UP and get retrained in other fields, get themselves healthy with clean air and water, and STOP their complaining about something that they cannot change.

PS. "coal-state Democrats retreated on long-term health care benefits for retired miners but promised a renewed fight for the working class next year."

The Coal mining industry has over 110 years of warnings and 110 years of ignorance and violations, it did not heed the warnings; what did people think would eventually happen?

Close the coal mines and stop the mountain and environmental damage as well as the damage to the workers and their families. There are alternative energy sources that are much less harmful, and frankly the Coal Industry does not have that many workers any more that we should worry about a major unemployment problem, these people can be retrained, presented with alternative industry opportunities, and end up with a happier, safer, and healthier life. Perhaps the solution is to just pay for their retirement at any age, and save us from the hazards of the industry that are costing billions.

What have we gained from the Coal Industry?

* Black Lung causes death and a drain on our medical insurance and Medicare, and Medicaid.

* Hundreds of mountaintops removed, causing destruction of nature, interruption of habitat, and environmental problems.

* Towns destroyed by explosions, fires, and sludge floods.

* Rivers and the ecosystems destroyed by sludge floods.

* Train wrecks and spills due to the heavy weight of the coal loads

* Acid rain that destroys car finishes and building roofs, as well as being an environmental disaster to trees and other plant life.

* Air pollution that is causing cancers, lung diseases, and just plain bad breathing in many states.

Oil Production:
"As the demand in China, India, and other countries goes up, the price of oil will rise." REF: www.SeniorCitizenLocalWeb.com

"Do we have the technology to convert our ships, planes, lawn mowers, power plants, SUVs, 18-wheelers, trains, cars, busses, motorcycles, heating systems, factories, military and other oil dependent tools to corn oil or other Green Energy Fuels? Can the average homeowner afford to replace each of his or her toys and heating/cooling systems in the next year or so? The answer to each is 'NO' we do not and cannot." REF: www.SeniorCitizenLocalWeb.com

"There will eventually be four major world trade organizations, the USA with Canada, Mexico, Central and South America; the Arab lands, the European Union, and the Russian, India, China, Japan cartel. This will

pit one another against each other in a battle for oil, water, food, and other resources, all of which are shrinking daily. We in the USA need to rethink our ways and lifestyle, and get ready for it." REF: www.SeniorCitizenLocalWeb.com

"OPEC has been very cooperative with the USA and others in the world and has held prices and production in check, it is the speculative nature of oil futures caused by outside forces that continue to drive up the prices." REF: www.SeniorCitizenLocalWeb.com

"Work from home as much as possible. LOVE TO. The problem is that we do not have employers that trust people to do the job for which one is being paid. This is just one way in which we must rethink our way of life and make the change. Writers, programmers, accountants, engineers, and all other pencil {keyboard} pushing jobs can and should be done from the home office; this would save us billions of man-hours, billions of gallons of gasoline, etc." REF: www.SeniorCitizenLocalWeb.com

What we must consider is that the burning of oil is much the same as the burning of coal; it produces the same toxic poisons. Burnt oil, i.e. gasoline, and other petroleum products produce harmful toxins that pollute the air we breathe and the water we drink. Additionally, burning hydrocarbons adds to the particulate matter that is covering the earth and being consumed by plants and animals, the very same plants and animals we tend to eat.

Manufacturing Pollution:

When one is manufacturing certain items he or she many times need to use corrosive chemicals, oils, paints, plastics, and other items that have proven to be detrimental to the planet. Although these items in many instances cannot be replaced with less harmful items, some may. The key is for companies to be educated via schooling or heavy fines in the proper handling of these materials, the proper disposal of the materials, and the proper substitution of materials.

12

The manufacturing companies need to be responsible and stop polluting and destroying to the point that the Taxpayers have to pick up the 'clean-up' bills (Superfund expenses), the medical expenses, and the cost of losing valuable land, water, and air.

We have regulations to 'prevent' greed from taking over the business community, and if they behaved themselves and acted like responsible citizens we would NOT NEED Regulations, but until then.

For those readers that do not know what a 'Superfund site' is, it is a place where irresponsible companies dumped materials that may have caused cancer, lung problems, deformed births, shortened life spans, and death to humans and animals. Sites may contain poisonous Lead, Mercury, Arsenic, and any of hundreds of other contaminants that eventually leach into our ground waters, our rivers, our lakes, our bays, and our oceans. There are over a thirteen hundred Superfund sites in the US, and the cost of cleanup is into the tens of billions of taxpayer dollars.

Chemical Production Pollution:

Been to the doctor lately? Did he or she prescribe some medications? Did you take these medications? Your body uses chemicals (Medications) for healing itself from a variety of ailments. The problem is that you pass these chemicals as you urinate or defecate and the chemicals end up in our water systems. Our sewage plants can test for and filter out certain elements but there are just too many, or it is just too expensive to filter out all. Thus, each glass of water we drink may have a 'trace' amount of someone else's medications that may cause mutations to us or to the marine lives that resides in our water.

We also create chemicals for washing our dishes and clothes, washing our bodies and our cars, fertilizing our plants, killing bugs and critters like rats and mice. Each one of these chemicals eventually get into the water supply and our oceans, and each contribute to the 'Dead Zones' that we are experiencing across the globe.

Nuclear Pollution:

This author worked for a contractor that contracted to the Atomic Energy Agency in the testing of Atomic Weapons at Mercury Nevada in the 1960s. We exploded Atomic Bombs and then sampled the dust and air for radiation and other chemical contaminants. The results were that the air and the particulate were contaminated with radiation that can cause cancer and genetic mutations for decades after the blast.

There has also been the Three Mile Island incident in the US, the Chernobyl incident in Russia, and the 2011 Fukushima Daiichi nuclear disaster in Japan where contaminated items are still showing up on the west coast of the US. Over the years we have had many civilian and military accidents that have put dangerous radiation into our air and waters.

REF:
wikipedia.org/wiki/Lists_of_nuclear_disasters_and_radioactive_incidents

Shipping Pollution:

One might question this, but it is true. Every time we package an item for shipping and distribution we use cardboards (boxes), plastics (packing material), inks (labels and logos), paper (labels), and glues (sealers) that eventually get thrown into the garbage dumps. These materials, with the possible exception of the plastics, dissolve when sun and rain saturate each, and the chemicals that make up the materials leach into the ground and eventually our water supplies.

Trash Pollution:

The Great Pacific Garbage Patch brings us back to the 1965 Robert D. Conrad story where we found trash over a thousand miles from land. Only this time we are in the middle of the Pacific Ocean and the pile of floating trash and plastic are being formed by 'Gyres' a vortex like flow that pulls anything floating into it.

System 001 is attempting to clean up the estimated 142,000 tons of plastic garbage that floats on approximately a million square miles of the Pacific.

REF: usatoday.com/story/news/2018/11/02/ocean-cleanup-system-begins-tidying-up-great-pacific-garbage-patch/1774506002/,

One of the problems with the floating garbage is that there may be tons of microplastics, i.e. plastics that were thought to be safe as each large size item over time breaks into very small particles. The problem is that these small particles can be found many feet to miles below the surface of the water, and can be swallowed by many marine species. We may then catch and eat these and thus we end up eating plastics that may be detrimental to our health.

Another problem with all the trash floating out there is that many marine lives tend to get caught in the junk (Usually discarded fishing nets, or plastic bags and six-pack holders); others feed on and choke to death on items that they cannot swallow or digest.

REF: marinedebris.noaa.gov/info/patch.html

Section - # 04 – Dead Zones:

There are several 'dead zones' around the world. These are patches of ocean that are so saturated with algae and chemicals that no plants or life can exist. These 'zones' are getting bigger each day we keep using fertilizers and pesticides on our lawns, fields, homes, streets, etc.

Streets? Yep, back to the farm in the Catskills. I said that we made candy, and canned jams and jellies. The rural country roads back in the 1940/1950 eras were lined with Strawberries, Elderberries, Blackberries, Red Raspberries, and other edibles like Horseradish, wild Onion, Rhubarb, Dandelion, Walnut, Butternut, and Cherry trees. The highway department would come out once a year and cut the majority of these so that one could better see the edge of the road and the drainage ditches. After a while, about in the 1960s they got lazy and decided to just 'spray' the ditches and road edges with 'killers' that eliminated all grass and plants. The rains then carried this poison into the nearby stream, and the steam to the Wallkill River, and the Wallkill to the Hudson, and the Hudson to the Atlantic. Total destruction of the plants, of the local fish, and of the recreational use of beaches along the lower Hudson and the bay entering the Atlantic.

Eutrophication:
Eutrophication is where the amount of oxygen in the water is too low to support life, and the elements Phosphorus and Nitrogen increase; thus allowing algae to bloom out of control. As Nitrogen increases so does Oxygen, but the sheer amount of the Nitrogen eventually kills the algae, and decomposing algae uses up all the Oxygen until no Oxygen remains. Without the Oxygen all marine life dies in the 'zone'.

Gulf of Mexico:

"100K gallons of oil NATURALLY seeps into the gulf each day..."
{Statement made after the British Petroleum Disaster and Spill}

What you are not seeing is the story about how we have thousands of old, up to and over 100 years old, wells in the gulf that are or will be leaking poisons for tens of hundreds of years. Many of the older wells were capped with concrete that is slowly dissolving into the water and will eventually, if not already have start leaking. Also the wells have IRON liners, pipes that do eventually rust and fall apart, this too is causing crude oil to leak into the gulf. What the oil companies and those that back then do not want you to know is that this disaster CANNOT BE STOPPED, the Gulf of Mexico is Dead, and WE DID IT.

> "The gulf is nearly back to pre spill condition"
> {Statement made after the British Petroleum Disaster and Spill}

Let's see, Back to Normal. Thousands of birds dying, hundreds of sea turtles washing up dead or sick, hundreds of whales and dolphins washing up dead or sick, hundreds of 'baby' dolphins washing up dead on the beaches, and miles of acreage covered with foot deep crude oil from state border to state border; sounds like a plan to me. REF: AP (Associated Press)

Mediterranean Sea:

There are areas along the shores of Greece and Italy that have now restricted fishing due to both over fishing, and the pollution that makes for near dead zones (Keeping people out may help the zones recover) where oxygen is low and phosphate pollution is high.

> "The United Nations Environment Programme has estimated that 650,000,000 t (720,000,000 short tons) of sewage, 129,000 t (142,000 short tons) of mineral oil, 60,000 t (66,000 short tons) of mercury, 3,800 t (4,200 short tons) of lead and 36,000 t (40,000 short tons) of phosphates are dumped into the Mediterranean each year."

REF: Explorecrete.com

> "The discharge of chemical tank washings and oily wastes also represent a significant source of marine pollution. The Mediterranean Sea constitutes 0.7% of the global water surface and yet receives 17% of global marine oil pollution. It is estimated that every year between 100,000 t (98,000 long tons) and 150,000

t (150,000 long tons) of crude oil are deliberately released into the sea from shipping activities."

REF: wikipedia.org/wiki/Mediterranean_Sea#Environmental_issues

Chesapeake Bay:

The bay is polluted by the water runoff of poultry farms that produce large amounts of chicken manure and by shoreline factories that produce airborne Nitrogen that is absorbed into the water.

Baltic Sea:

"Dead zones have increased by more than 10-fold in the last century - Baltic Nest Institute".

REF: balticnest.org. 2014-04-01

There are several other 'Dead Zones' around the world, including part of the Great Lakes, the Saint Lawrence River, and the coastal area of Oregon. Some areas have been cleaned up, the Black Sea being one, but by accident. When the Soviet Union was dissolved, the use of Nitrogen rich fertilizers became too expensive and thus the use of it came to a halt, and that allowed the Black Sea via fresh water entry to 'heal'.

REF: wikipedia.org/wiki/Dead_zone_%28ecology%29#cite_note-nationalgeographicbay-20

Section - # 05 – Cost to Individuals:

I no longer live along the east or west coast and therefore do not have to pay for the cleanup of ocean storms and floods. Right? Nope, Wrong!

We all chip in to the fund we call 'Taxes' that is used to help solve problems throughout the US and in some instances the world.

Fires, Floods, and Insurance:

Over the last decade we have had several major hurricanes, tornadoes, fires, and floods that have cost hundreds of billions in property damage and loss of life. In some of these instances the personal and homeowner's insurance covered the losses; in most instances the taxpayers via FEMA (Federal Emergency Management Agency) is the funding mechanism that rescues people from total bankruptcy and personal losses.

After hurricane Sandy many insurance companies raised premium rates for those living along the Gulf of Mexico and the East Coast of the US. Some companies now refuse to even issue policies; the potential losses are just too steep. As an example:

> *"Florida has some of the highest rates for homeowners insurance in the country – on average $1,918 per year for an HO-3 policy, compared to the nationwide average of $1,192. That said, how much you'll pay can vary a lot depending on your home's size, your assets, and your address."*

> *"To find good homeowners insurance in Florida, you'll need to get to know some insurance companies you might not recognize – **many of the big insurance providers avoid Florida**, and the state's top six providers (by market share) include a few local insurers and one government option."*

REF: Reviews.com

In California we have not floods, but fire due to long-term weather changes. This has resulted in many areas not being able to get fire insurance, and many areas where the premiums have nearly doubled.

What has this to do with Global Warming? Well, as the oceans temperature warms up there is more evaporation, and this adds to the amount of water vapor in the sky. During a storm of any kind this increase in water vapor rains down upon the land and causes floods that can damage property; this is exaggerated many times during a hurricane like Sandy, Katrina, and Dorian. Also, a major storm that is on the ocean produces waves that can come ashore and flood towns like Charleston, North Carolina, and if there are overflowing rivers trying to empty into the ocean's tidal and storm currents, the height of the water can reach dozens of feet and thus destroy vehicles, homes, stores, and everything else in its path.

Heavy rain also can cause hillsides to slip and cover homes, roads, and streams, thus causing much expense for the cleanup and rebuilding of the destroyed properties.

Heavy rain can also result in a bumper crop of ground cover, i.e., weeds, grasses, small trees, brush, etc., and this ground cover when dry contributes to fires that are set by man or nature. California lost thousands of homes due to excessive ground cover fires.

Home and Business Destruction:

Global Warming is adding to the strength of storms, not necessarily the frequency of each. Most new homes are built to the National Building Codes, the National Electrical Codes, and the National

Plumbing Codes and therefore, can withstand up to about 100 MPH (Miles per Hour) winds. Older homes and homes that are trailers or shacks are not built as strong and can be destroyed by lesser winds.

We have seen this on the islands of Puerto Rico (2017) and now the Bahamas Islands (2019), and the amount of destruction is great as many thousands of homes and businesses were totally destroyed, thousands of lives lost, and island economies devastated.

Storms like Dorian that have sustained winds of 220 MPH are nearly impossible to design for, but with the proper technology we can do it. The prime problem is the cost of the designs and construction methods that are financially out of reach for most of our citizens.

To the insurance companies, the local towns, and the taxpayers the cost of a storm can include the 'prep' for the storm, the security required during and after the storm, the removal of debris after the storm, and the cost of rescuing or treating people that failed to leave when ordered, or who failed to even know the dangers. This cost can be into the tens of millions, if not billions.

For businesses, the loss of business and the time to inventory, restock, gain back customers, etc., can be days to weeks and many businesses never recover.

Farmland Destruction:

What if, what if you owned a farm that produced a crop that required 120 days of 60 to 85 degree weather, a rain every few days, and good topsoil that was full of nutrients. The farm provides a decent crop that can be sold for a profit and you and yours are living the good life.

Suddenly, the weather pattern changes and the 60 to 85 degree temperatures are now 80 to 105 degrees, the rain only falls for a few minutes every week or two, and the topsoil starts to blow away. How would you feel and what would you do?

Sounds like a 'never can happen' scenario, but it did happen. In the 1930s in the Midwest we had forest, abundant trees and lots of small homesteads with a garden and maybe a field or two. We got greedy and decided we wanted more, thus we cut down the trees, plowed up the land, and planted tons of wheat and other crops. The lack of trees to provide cover (Shade and moisture), and the plowing up of the ground (it is now prone to being blown by the wind) resulted in what is known as the 'Dust Bowl', a period of our US history that should never again be allowed.

{This year (2019) we are watching the Amazon Rainforest be destroyed in the same way, and its destruction will result in a catastrophe}

Today we have the opposite; we have major storms that dump millions of tons of water on our fields and into our rivers. This results in floods that drown the land and thus, kill off the crops that are planted. The recent (2017 & 2019) Mississippi River floods did considerable damage for thousands of square miles of farmland. It did though, replenish much of the soil and in the long run may be a 'blessing' to the future of farming in the regions, that is if no more floods happen and the newly placed topsoil is not too polluted with poisonous chemicals that washed off the lawns, streets, and fields far upstream. These floods are the result of drainage plains of the Missouri, Ohio, and Mississippi Rivers experiencing massive rains.

The floods were the result of excess water vapor in the atmosphere due to the warming of the Gulf of Mexico and the Pacific Ocean. The results were higher than normal rain storms, more runoff, and the destruction of crops, vehicles, farm equipment, towns, stores, homes, schools, etc, that lined the banks of the Missouri, Ohio, and Mississippi Rivers.

Secondary to the floods is that millions of tons of pesticides, Nitrate fertilizers, and other chemicals flowed south to the Gulf of Mexico further amplifying the already massive Gulf of Mexico 'Dead Zone'.

Transportation Delays:

Every time we have a major storm we have transportation delays; roads and bridges close, or there are tremendous backups as thousands to millions try to evacuate low lying flood prone regions. Roads and in some cases airports and train tracks that are underwater are not safe for travel and it can take from hours to days for the water, muck, and debris to be cleared. This can result in tourist trade slowdowns; products and materials not getting to market of manufacturing facilities, and infrastructure destruction that may not show up for days, weeks, or even years.

Additionally, when massive storms are forecast there are usually an exodus of people at the airports, and during the storms many airlines will suspend operations, which is an economic hardship to both the airlines and their passengers.

What many do not understand is that there is also a safety concern when the weather gets too hot. Aircraft need air under their wings in order to take off and fly, and if the air is too hot, as it is in cities like Phoenix Arizona at times, aircraft cannot take off, the air will not support it.

Recreational and touring facilities like Hot-Air Balloon rides come to a halt as each too cannot fly.

Infrastructure Destruction:

Climate change via Global Warming can cause heat that buckles pavements; rain that flooded roads and underground railroad tracks (NYC subways saw this during Sandy). Excessive water can undermine foundations, roads, track beds, bridges, dams, and destroy piers and shipping channels. It can also destroy beaches and thus the recreational facilities that support many of our oceanfront towns.

The wind from storms take out power lines and buildings that may be part of the police, fire, and military infrastructure; and the downed trees can close roads that are needed for recovery after a storm.

Fish and Shellfish Production:

The change in ocean temperatures is affecting the habitats of our marine life and as the water warms due to Global Warming, many species are dying off or moving to cooler waters. There is also the factor of melting ice at the caps, in Greenland, and on our mountains (Glacier National Park is almost out of Glaciers). Ice is fresh water, and the massive amounts of it melting into the sea are in some local areas diluting the salt and other minerals that seawater contains. This also affects the marine life and the texture and taste of some of the marine products we enjoy eating.

Ocean acidification
Ocean acidification is detrimental to shellfish like clams and oysters and with the melting of the glaciers the chemical balance of the water is changing and becoming more acid in value. This prevents the shellfish from growing strong shells, and thus produces an inferior product. The acid level also affects 'Terapods' that are the main diet of young pink salmon.

REF: https://www.kcaw.org/2012/11/20/how-do-melting-glaciers-change-ocean-chemistry/

The Blob is Killing Fish:

In the Pacific the ocean has for a second or third time heating up by as much as 7 degrees. This 'Warm Blob' is dangerous as it produces an El Niño that can greatly affect the weather in California and the southwest. Marine life is also affected as follows:

REF:
https://www.forbes.com/sites/allenelizabeth/2019/09/05/another-warm-blob-is-forming-in-the-pacific-ocean/#2dc719f314af

Tax Dollars:

When we have a major storm we have to secure property, guide people to the shelters, take care of people, rescue those that failed to heed the warnings, clean up the streets and roads, and much more. Every bit of this cost money, either at the local, country, state, city, or federal level and that money comes from 'tax dollars'. You may not believe in Climate Change, or Global Warming, but you can bet your hard earned money is going for cleaning up the messes left behind by these massive fires, floods, and storms.

Section - # 06 – Military Concerns:

Our military consists of the US Army, US Air Force, US Marines, US Navy, the Coast Guard, and our National Guard and its Reserves. During major storms and fires many of these fine folk are called upon to aid in the protection, rescue, recovery, and rebuilding of people's lives and property. Our US Army Corp of Engineers are the builders of the dikes, flood control dams, and much more. Areas like the low areas of New Orleans should not be used for housing, it cost all of our taxpayers for the engineering and construction of the dams and dikes so that some of our citizens can continue to build and rebuild and rebuild after each flood.

Military Considerations:

Our military are becoming worried as the oceans rise, the storms get progressively worse, and it gets more difficult to protect many of our bases from the weather. Think about the cost of a storm to the military that has to move dozens of ships from the storm's path; hundreds of aircraft to safer airfields, and then 'button down' all the buildings on each base that may be hit by the storms. This cost is in the tens of millions for each storm.

REF: https://www.militarytimes.com/news/your-military/2017/09/12/pentagon-is-still-preparing-for-global-warming-even-though-trump-said-to-stop/

Population Considerations:

This past week (August/September 2019) we saw Hurricane Dorian nearly totally destroy the Bahamas and then wreak havoc on our east coast communities. Across the planet the oceans are getting angry, the warming of the oceans expands the water and it therefore rises in levels. The warming of the air is melting ice caps and glaciers that are also raising water levels. The combination of these two factors are causing what once were 'safe' land, i.e. shores and islands, to be either flood prone or actually underwater. These in turn are causing millions of coastal and island dwellers to seek higher and dryer ground and in doing so these people are seeking help.

Who are the helpers? Is it to be the USA or will it be the Russians, Chinese, or some other nation. Think about it, do we really want to have a Chinese owned island, i.e. the Bahamas or Puerto Rico just miles from our mainland? If not, then we need to provide the helpful services that are needed for economic and social recovery of these islands before, during, and after a massive and destructive event like Hurricane Dorian. And the best people we can send and use are our military that is trained for these types of events.

Section - # 07 – Environmental Laws:

President Jimmy Carter (D-GA) was one of the first leaders to attempt to inform and change American minds about our contribution to Global Warming and Climate Change. He even had Solar Panels installed on the White House that were immediate removed by President Reagan (R-CA) upon his taking office.

President Richard M. Nixon (R-CA) created the organization we know as the EPA or Environmental Protection Agency (December 2, 1970) that is currently being disassembled by President Trump (R-NY).

Green Energy:

One passenger train can easily hold a few hundred people, which means that it can eliminate dozens of passenger cars from our highways. Railroads built this nation and unfortunately the owners of many of our railroads, like the owners of our steel and communications industries, decided to 'keep' the profits and not invest back into the product that made them rich in the first place. We lost most of our passenger railroads, and thus the manufacturing plants that made the track, made the engines, made the passenger cars, and built the stations. This eliminated tens of thousands of good paying jobs as we 'gifted' our newer technologies to Japan, Europe, India, and China; all of which do manufacture high-speed rail systems that have cut dirty energy use.

Green Job Information:

8:53 AM 5/29/2012

> In 2011, the percentage of total employment associated with the production of Green Goods and Services (GGS) increased by 0.1 percentage point to 2.6 percent, the U.S. Bureau of Labor Statistics reported today. The number of GGS jobs increased by 157,746 to 3,401,279. GGS employment accounted for 2.3 percent of private sector jobs and 4.2 percent of public sector jobs in 2011. The private sector had 2,515,200 GGS jobs, while the public sector had 886,080 GGS jobs. Among private sector industries, construction

had the largest employment rate increase, from 7.0 to 8.9 percentage points, while manufacturing had the most GGS jobs (507,168). (See table 1.) GGS jobs are found in businesses that primarily produce goods and provide services that benefit the environment or conserve natural resources.

REF: www.bls.gov/news.release/pdf/ggqcew.pdf

Green Jobs Include:

First there are the people that create the ideas, then those that do the engineering, followed by those that find the money, and then the people that do the manufacturing and construction. Making things 'Green' or energy efficient can take hundreds to thousands of well-paid workers. We are not just talking about Wind Turbines, or Solar Cells, but also better built homes and commercial buildings; more energy efficient lighting, cooling, and heating; gasoline free transportation; high-speed rail systems; more efficient farms that use its waste for energy production; the manufacture of non-polluting products, and much more.

Section - # 08 – Damages Caused by Not Going Green:

Endangered Species:

In the last 40 years we have managed to kill off about 50% of the species that existed on our earth for centuries. Our survival has created a situation where we are eliminating the habitat of creatures each and every day, and the outlook for the future is not good. Many of our medicines, foods, and other items come from species that many of us have probably never seen or encountered in our daily lives, but we do depend on without knowing it. And in nature there may be that one small animal that is living on a deserted patch of land that may be the animal that has the cure for all Cancers, or Alzheimer's, or some other disease that can end civilization. Do we really want to 'kill' it off before we make that discovery? I believe not!

REF: https://naturalenergyhub.com/pollution/endangered-species-due-pollution/

Acid Rain in 2012:

Oil and Coal burning smog are causing Acid Rains (Sulphur Dioxide, Nitrogen Oxides, and Water) which are ruining roofs, car paint jobs, trees and plants, rivers, lakes and oceans, and according to thousands of scientists are causing Global Warming, and according to thousands of environmentalist pollution is contributing to the death or illnesses of animals, birds, insects, and fish world wide; this folks is our food supply.

So, you live in your polluted world and enjoy your future Lung Cancer, Lead poisoned fish for dinner, Mercury poisoned seafood at your favorite restaurant, and swimming in acid that rots your skin, I do NOT want to Live in YOUR world.

I prefer to use the energy of the sun, the energy of the earth, the energy of the wind and water flows, and I prefer to spend my money on energy efficient appliances, energy efficient transportation like rail, energy efficient insulation and doors and windows, natural home grown foods, and both passive and active means of heating, cooling, and enjoying life and it benefits on this planet.

REF: https://www.conserve-energy-future.com/causes-and-effects-of-acid-rain.php

Oil Spills:

Ever work on a car or do some tar work on a roof? The oil, grease, and tar products are all petroleum products and are sticky and very messy. If you get the stuff on your clothing, tools, or hands it becomes a difficult situation as you try to remove it. Degreasers, soaps, and other chemicals are needed and these all harm the environment if not properly used. Your hands have a surface area of about 90 square inches total, and it takes considerable cleaner and lots of effort to get each clean.

Think about the effort and amounts of 'chemical cleaners' that are needed to clean hundreds of square miles of beach. And think about the damage to the water and the marine animals that are first harmed by the spilled oil, and then by the cleaners used for removing the spilled oil.

Biodegradable vs. Non-Biodegradable:
Which do you think is a better product, a disposable plate made from paper (wood fiber), plastic (Petroleum) or Bagasse (Sugar Cane fiber byproduct)?

All three items will eventually degrade into a semi-harmless substance, but wood if not from already used wood (like the remains of a house after a storm) is harmful to our air; plastics can last for a hundred years; the Bagasse for two years or less.

We have cleaners that are considered biodegradable, and we have many that are not. Both are harmful to our environment, but the amount and length of harm differs greatly. Do your homework and carefully select cleaners and products that are biodegradable, your planet will thank you.

For a listing of environmentally friendly, and unfriendly cleaning products Google™ "Environmental Working Group"

Capped Wells are Failing:
BP, Mariner, are wake up calls to AMERICA and AMERICANS; we are getting into deep trouble and the self-policing industry is doing little to solve the problems it is creating. Capped wells are failing due to very old and rusty metal, cement that is falling apart, earth cracks, etc., and oil leakage in the Gulf is increasing daily, and will

eventually destroy the entire ecosystem. More rigs being built closer and closer together will hamper shipping and eventually we will have an out of control ship ram a rig, this fear prevented the 'Whale' from operating in the BP disaster.

The oil spills also have an effect on the wildlife that depends on clean waters and set PH values of the waters. The Gulf of Mexico has thousands of very old and leaking wells, but this is not published by our news media in fear of 'hurting' the 'advertisers', like BP, EXXON, GULF, MOBIL, etc. These wells date back to nearly 1900 and although many have been capped, the technology of the period did not measure up, and therefore the cements and steels are failing. We will someday see a 'dead sea' off our southern coast. Now the oil companies want to 'repeat' the crime in the cold arctic waters off the north coast of Alaska, where men and machines cannot easily go OR provide necessary cleanup in the event of another drilling rig disaster.

Hurricane Storm Damage:
Hurricanes are getting stronger with more rain and stronger winds, and these winds have damaged oil rigs, destroyed oil pipelines, and polluted the gasoline in storage tanks. The cost of this runs into the billions over the decades and that money comes directly out of your pocket as increases in prices for plastics, highways, and fuels.

Oil Spills:
Over the decades we have had oil spills in Alaska, the Gulf of Mexico, in several foreign nations, and across the US as pipelines age. The average oil pipeline has a life expectancy of 42 years, and the major of our pipelines are either approaching that age or have long exceeded that age.

Oil spills like the Exxon Valdez, the Mariner Energy Gulf Oil Fire, British Petroleum, and the hundreds of others not only disrupt the flow of commerce, but also cost the consumer in the long run. The cost includes not just the cleanup, but also the loss of seashore businesses, the seafood production industries, the recreational tourist industries, and the loss of environment that keeps our earth in balance.

Section - # 09 – Letters from States Persons:

I spent two years creating a southern border system for securing our border, protecting wildlife, protecting farmers along the border, creating thousands of current and millions of future jobs, and providing near unlimited Green Energy to the Southwest.

See this author's Amazon.com/Kindle book and *"BorderTransportationSystem.com"* website for tons of details.

When finished the suggestions were submitted to some of our congress, to the presidents of the US and Mexico, and to the NPS (Department of the Interior). The following are a sampling of the responses received.

Green Jobs by Allyson Y. Schwartz

I wrote to Congresswoman Schwartz about my concern about our failure to insist that are government push for Green Energy and Green Energy Jobs. The following is her response as of 4:49 PM 4/4/2012

Dear Friend,

This week, I had the opportunity to address the Good Jobs, Green Jobs East conference, which was held right here in Philadelphia. This organization, now in its 5th year, holds regional conferences to move the country 'forward, to a cleaner and more prosperous economy'.

During my speech, I explained that I firmly believe economic growth and environmental stewardship are not only compatible, but are immeasurably linked. In fact, ignoring the need to develop renewable energy, failing to use valuable resources in environmentally sustainable ways and neglecting the potential for energy conservation, means a missed opportunity for jobs here at home, for energy savings and for greater energy independence.

Today, 3.1 million Americans are working in "green jobs." They are weatherizing homes; retro-fitting businesses; producing and installing solar panels and windmills; designing advanced batteries; researching and developing more energy efficient everything from buildings, to appliances, from furnaces to automobiles; and developing new sources of clean energy; and new ways to recycle and reuse industrial and residential materials.

The Philadelphia Inquirer reported on March 23, of the 3.1 million green jobs, "182,193 were in Pennsylvania, making it one of the five highest states in the country in generating green employment." Read the full story here.

I have long been an advocate for clean energy jobs and sustainable communities, believing that while they may call for some changes in the way we do things, these changes will produce jobs, reduce energy costs and make us more self-sufficient.

As part of an effort to promote energy efficient commercial buildings, I was successful in adding an increase in the tax credits for energy efficient commercial buildings that achieve 50 percent greater efficiency. This tax credit is being put to good use in construction projects across the country. I have been working to modernize the Historic Tax Credit, in part to promote energy efficiency as we restore and reuse older buildings in cities and towns across the country. I am also working to promote the greening of local communities in urban and rural America through local community planning.

I have also authored legislation to offer tax incentives to startup companies to stimulate the development of biofuels. And I am working closely with the Philadelphia Water Department on their application to the EPA to reconstruct the aging water system here in Philadelphia, through an extensive combination of green and gray infrastructure. This exciting and remarkable effort could be a huge milestone for our country.

Unfortunately, Congressional Republicans have focused on weakening environmental regulations and subsidizing fossil fuels at the expense of clean energy. They insist on maintaining $40 billion in tax credits for the largest, most profitable oil and gas companies.

But I am not deterred. There is no better time to put Americans back to work by building the greener, cleaner economy of the future then right now.

TAKE MY POLL ABOUT GREEN JOBS

As always, it is a privilege to represent you in Congress.

Sincerely,
Allyson Y. Schwartz
Member of Congress

PA Biotechnology

In 2011 I wrote Congresswoman Schwartz about the possible use of Biofuels and Biotechnology for replacing our coal, oil, and natural gas with clean energy sources. Here is her answer.

Dear Friend,

As our economy continues to recover, it is important that we create opportunities that promote private sector economic growth. In Pennsylvania, biotechnology is a vital sector of our innovation economy and is ripe for growth and job creation.

In Philadelphia alone, the life sciences sector is responsible for creating one out of every six jobs, and generates 15 percent of all economic activity. By investing in the life sciences field, we are providing universities, research centers, and private companies with the tools they need to create good paying jobs and be on the cutting edge of medical advances.

That is why I have been working hard to promote opportunity in this critical economic field. This week I introduced bipartisan legislation to provide tax incentives for small and mid-sized businesses to invest in life sciences research and development on a targeted basis. This bill will grow our economy and increase America's global competitiveness by enhancing medical innovation, life sciences education, and job creation here in the U.S.

The Life Sciences Jobs and Investment Act encourages investment in the United States by reducing the tax burden on the mostly small and medium sized companies responsible for life science research today. The bill allows companies engaged in life sciences research to either double their Research and Development tax credit on the first $150 million invested or repatriate foreign earnings tax free up to that same limit when used exclusively for job creation and research in the United States.

Investing in life sciences means:

• Hiring more scientists, researchers, and personnel to engage in new and groundbreaking research;

• Allowing American universities and post-graduate institutions to allocate more time and support for research and development; and

• Investing in new laboratory and life science research facilities and equipment.
We must continue to invest in major medical innovation for the future and ensure that America remains on the cutting edge of innovation and scientific research.

The new legislation builds on my earlier success with the therapeutic tax credit, which invested $1 billion in nearly 3,000 small biotech and bioscience companies in 2009. This program created high-skilled, high-wage jobs in Pennsylvania and throughout the United States.

The Life Sciences Jobs and Investment Act will expand these efforts and continue to propel job growth in the Commonwealth. Both of these bills were introduced with strong bipartisan support.

Click here for more information on the Life Sciences Jobs and Investment Act.

Click here for more information on the Therapeutic Tax Credit Extension.

If you would like more information on the tax credit, please call Brandon Casey in my Washington, DC office at 202-225-6111.

Sincerely,
Congresswoman Allyson Schwartz

American Renewable Energy Production Tax Credit Extension:

8:06 AM 1/27/2012

I wrote to Senator Robert P. Casey, Jr. about H.R. 3307 a tax credit extension act. Here is the response:

Response from Senator Casey   From: Senator Robert P. Casey, Jr. <senator@casey.senate.gov> Date:Thu, Jan 26, 2012 6:43 pm

Dear Mr. Jamerson: {I used an AKA name}

Thank you for taking the time to contact me regarding H.R. 3307, the American Renewable Energy Production Tax Credit Extension Act of 2011. I appreciate hearing from you about this issue.

American Renewable Energy Production Tax Credit Extension Act
Representative David Reichert of Washington introduced H.R. 3307 on November 2, 2011. The American Renewable Energy Production Tax Credit Extension Act would amend the tax code to extend through 2016 the tax credit for electricity produced from wind, biomass, geothermal or solar energy, landfill gas, trash, hydropower, and marine and hydrokinetic renewable energy facilities. No similar bill has been introduced in the U.S. Senate.

Should this legislation reach the full Senate for consideration, please be assured that I will keep your views in mind.

As always, I appreciate your views, thoughts and concerns as they assist me in understanding what is important to the people of Pennsylvania. Please do not hesitate to contact me in the future about this or any other matter of importance to you.

If you have access to the Internet, I encourage you to visit my web site, http://casey.senate.gov. I invite you to use this online office as a comprehensive resource to stay up-to-date on my work in Washington, request assistance from my office or share with me your thoughts on the issues that matter most to you and to Pennsylvania.

Sincerely,

Bob Casey
United States Senator

Section - # 10 – Letters from Organizations:

This section contains responses from various organizations to inquiries that I made over the years and some quotes made on the Internet that are public domain.

Coal Companies Names and B of A:

2:43 PM 11/15/2011

"Arch Coal and Peabody Energy, two of the biggest coal mining companies in the Powder River Basin that are trying to turn the pristine Pacific Northwest coastline into a major hub for exporting coal around the world."
Source:
Amanda Starbuck
Energy & Finance Campaign
Twitter: @DirtyEnergy

"Edison International, which owns the old, dirty Fisk and Crawford Plants in urban Chicago. Pollution from coal plants like Fisk and Crawford cause health problems that kill 24,000 Americans every year."

Source:
Amanda Starbuck
Energy & Finance Campaign
Twitter: @DirtyEnergy

Tell Bank of America: Not with Our Money!

Dear Chris, {I used Chris as a screen name, AKA}

Two climbers have just hung a banner reading "Not with our money" from Bank of America's corporate headquarters in Charlotte, North Carolina. To make sure the bankers inside got the message, we locked down the entrances, too, bringing business-as-usual to a standstill.

Bank of America puts profits ahead of people and the planet. The bank is investing in dirty coal companies that are polluting our communities and cooking our climate while also foreclosing on Americans' homes and laying off thousands of workers. The same shortsighted thinking that led to our global economic crisis is being applied to BoA's investments that impact the environment.

But not with our money. Not any more.

You can get in on the action too: Tell Bank of America executives "Not with our money!" right now.

We've just issued a new campaign briefing showing that, despite claims that it considers the impacts of its investments on the environment and the climate, Bank of America is the largest underwriter of the U.S. coal industry, contributing $4.3 billion to the coal sector over the past two years.

Bank of America invests in every dirty aspect of the coal industry, too, including loans to Arch Coal and Peabody Energy, two of the biggest coal mining companies in the Powder River Basin that are trying to turn the pristine Pacific Northwest coastline into a major hub for exporting coal around the world. BoA is also invested in companies like Edison International, which owns the old, dirty Fisk and Crawford Plants in urban Chicago. Pollution from coal plants like Fisk and Crawford cause health problems that kill 24,000 Americans every year.

We live in a time when the twin opportunities of job creation and the transition to a green economy are not only within reach, but desperately needed. Yet Bank of America, more than any other bank, continues to prop up coal, a dirty, 19th-century energy source.

Write to BoA now to say "Not with our money."

For a clean energy future,
Amanda Starbuck
Energy & Finance Campaign
Twitter: @DirtyEnergy

Die From Coal:

- 3:13 PM 4/7/2011
- Answer to a letter I wrote to Greenpeace.

Your and activists across the country made it clear to the Senate that the 34,000 Americans who die each year from coal and the extra half a trillion dollars it costs our economy is a price too high. The fight for clean energy continues. In Washington, coal industry lobbyists are pressuring lawmakers to strip the Environmental Protection Agency of its power to regulate the dirty emissions of coal-fired power plants. Our campaign is just getting started and we are going to need you every step of the way. Thanks for taking action and I look forward to working with you in the future.

I recently delivered your petition signature along with a copy of the latest Harvard study that shows the true cost of coal to both of your Senators.

You and 50,000 other activists signed our petition urging the Senate to defend the EPA's ability under the Clean Air Act to protect Americans from the true cost of coal. The Senate got your message.

During our delivery, we made sure to prioritize the offices of Senators whose states have been devastated by coal mining such as John Rockefeller (D-WV) and Mitch McConnell (R-KY). Also, those who take their cues from the polluter companies that paid for their election campaigns such as Lisa Murkowski (R-AK) and Jim Inhofe (R-OK).

We then dropped off the petition and the Harvard study to Senators like Sherrod Brown (D-OH) who have been on the fence about whether carbon pollution from coal is dangerous enough to restrict now.

Thank you for your support. With these petitions, You and activists across the country made it clear to the Senate that the 34,000 Americans who die each year from coal and the extra half a trillion dollars it costs our economy is a price too high.

The fight for clean energy continues. In Washington, coal industry lobbyists are pressuring lawmakers to strip the Environmental Protection Agency of its power to regulate the dirty emissions of coal-fired power plants. Our campaign is just getting started and we are going to need you every step of the way. Thanks for taking action and I look forward to working with you in the future.

Thank you,

Kyle Ash
Greenpeace Senior Legislative Representative

Mercury Poisoning and The EPA – 2011:

I wrote to Greenpeace to see what they had to say about the pollution situation, especially the Mercury in our air and especially our water supplies. Here is the response:

Subject: A toxin-free Mother's Day
Date: 5/5/2011 4:05:45 AM Eastern Daylight Time
From: webmaster@greenpeaceusa.org
Reply To:

To: csr@seniorcitizenlocalweb.com {website I created, no longer in use}

Sent from the Internet (Details)

As many as 1 in 6 American women have enough mercury in their bodies to put their baby at risk of brain damage, learning disabilities and birth defects.

And about HALF of the mercury found in the U.S. currently comes from coal-fired power plants.

That's why the Environmental Protection Agency (EPA) is going to limit mercury pollution coming from coal-fired power plants. But before the decision can be made final, one last round of public comments is required.

This is our chance to give EPA the public support necessary for this to hold up against the future attacks that will be coming from dirty energy interests, industry lobbyists and their friends in Congress once the decision has been made.

The comment period is only open for a short time. And given what's at stake for mothers, it seemed that Mother's Day was a great time to start collecting them.

This Mother's Day, give the gift of toxin-free air on behalf of the special mom in your life. Submit a comment to EPA now and urge them to strictly limit mercury and other toxic pollution from coal-fired power plants.

The facts are clear. Limiting mercury pollution will save lives. According to EPA, the proposed limits will prevent 17,000 premature deaths, 11,000 heart attacks and an astounding 120,000 cases of childhood asthma.

The moms, families and communities that live in the shadows of these coal plants are counting on EPA to stop toxic pollution. For these people, it is literally life or death.

And for the communities across the country who are working to shut down dirty coal plants, letting EPA do its job is absolutely critical. The coal industry knows this and that's why they are going to do everything in their power to roll back this decision. Every comment we collect makes it more difficult for them to do that.

EPA has taken a courageous stand against the industry by proposing to limit mercury pollution. Send them some love and send some love to mothers everywhere by submitting a comment today.

It's time to quit coal.

Happy Mother's Day,

From Credo Action / Credo Energy:

To: Allyson Y. Schwartz
Member of Congress

Subject: Let's ask them why they sold us out to Big Coal.

Dear Friend,

Talk about an inconvenient truth.

The coal, gas, oil, auto, and farming interests have proved much more powerful than environmentalists. Corporate polluters have managed to take decent legislation -- Reps. Waxman and Markey's climate bill -- and turn it upside down so that it will help them more than the planet.

It serves nobody to pretend that this is not true. Environmental organizations in recent weeks have sent dozens of requests asking their members to help strengthen and then pass this legislation. Ask any of theses groups if the strengthen part has happened on any of our demands. It has not.

I just signed a petition to ask my representative: "Why didn't you stand up to coal and strengthen the climate bill?" I hope you will, too. Please have a look and take action.

http://act.credoaction.com/campaign/climate_bill/?r_by=4695-283439-r2Eh_xx&rc=paste
Build a Clean Energy Economy - 2011

Union of Concerned Scientists (UCS)

I wrote the UCS about the use of vehicles and the EPA regulations that could reduce the harmful emissions from each. Here is the response:

2:08 PM 7/29/2011

From: Kevin Knobloch,
President of the Union of Concerned Scientists (UCS),

Dear Chris, {Middle Name, uses a AKA}

I just returned from a ceremony, convened by the White House, where President Obama unveiled an exciting new agreement with major automakers and the State of California to strengthen fuel efficiency and global warming emissions standards for new cars and trucks!

Exciting new fuel efficiency and auto pollution standards will save Americans money at the gas pump, curb global warming emissions, and reduce our oil dependence.

UCS has helped shaped this agreement since plans were announced by the president more than a year ago to develop new standards. I am so proud of the UCS clean vehicles technical, policy, and advocacy team – they unwaveringly kept their eye on the goal of deep global warming emissions reductions and insisted that the new fuel efficiency standards be ambitious and achievable.

This is a major step forward in our fight to make America's cars and trucks cleaner and more fuel-efficient.

For the second time since he was elected, President Obama has brought automakers to the table to support strong fuel economy and auto pollution standards for cars and trucks. Thanks to these efforts, we are on a path to dramatically reduce climate emissions from vehicles and make our country less reliant on oil.

This proposal, which will apply to vehicles sold in model years 2017 to 2025, will set a global warming pollution standard of 163 grams per mile by 2025 – the equivalent of 54.5 miles per gallon (mpg) if all improvements are met through fuel-efficiency.

Meeting these standards will unleash innovation in the auto industry, putting fuel-saving technology to work across all vehicle types and sizes. Automakers already have the technology to build cars, trucks, and SUVs that are less polluting and go farther on a gallon of gas. These standards will ensure more efficient engines, smarter transmissions, lightweight materials, and other clean vehicle technologies make it off the factory floor and into our driveways. It will also help put more hybrid-electric vehicles on the road and pave the way for electric-drive technology.

As a result, these standards will save Americans money at the gas pump, curb global warming emissions, and reduce our oil dependence.

According to UCS analysis, these new standards will cut U.S. oil consumption by as much as 1.5 million barrels per day by 2030 – about as much as we currently import from Saudi Arabia and Iraq.

This will prevent as much as 280 million metric tons of climate emissions in 2030 from being released into our atmosphere – the equivalent to shutting down 72 coal-fired power plants. By using less oil, Americans will spend $80 billion less on fuel.

This progress would not have been possible without the help of UCS and supporters like you. For years, UCS has helped lead the fight to strengthen fuel efficiency and global warming emissions standards for cars and trucks. This victory is no exception. We have worked in coalition with other public interest groups, communicated timely analysis to the media and key decision makers, and directly engaged policy makers.

Today's announcement is a giant step, yet important details must still be resolved before these standards are finalized next year. UCS will continue to work with the Obama administration, California, and other stakeholders to ensure these standards are strong and not weighed down by loopholes.

At a time when public health protections and climate science are under daily attack from some members of Congress, today's announcement shows that we can make progress even in this challenging political environment.

Thank you for your support and congratulations! I look forward to our continued work together to build a clean energy economy that tackles the threat of climate change.

Sincerely,

Kevin Knobloch,
President of the Union of Concerned Scientists (UCS),

{Note: In 2019 the Republicans attempted to rollback this legislation}

Greenhouse Gases:

"Greenhouse gases are accumulating in Earth's atmosphere as a result of human activities, causing surface air temperatures and subsurface ocean temperatures to rise."
-Climate Change Science, U.S. National Academy of Sciences, 2001

Letter from the President of Defenders of Wildlife.

Dear Friend,

This summer, a federal court in Alaska found that the Bush administration violated the law when it approved oil and gas leasing in Alaska's Chukchi sea without sufficient information and analysis about risks to the Arctic environment. A different court found similar flaws with Arctic drilling plans in 2009.

Yet, President Obama's administration has so far failed to revisit protections for the Chukchi – Arctic waters off Alaska's coast that are home to some of America's remaining polar bears and key to the survival of Inupiat Eskimo communities.

Please urge President Obama to take action now to save polar bears, bowhead whales and other wildlife – and the Alaska Native communities that rely on them to survive – by preventing the next offshore drilling disaster.

The Obama administration has asked the court in Alaska to allow activities that would pave the way for drilling to proceed, potentially jeopardizing an area key to the survival of not only polar bears, but also bowhead whales, Pacific walrus and other wildlife.

The Gulf oil disaster clearly demonstrates the terrible risks of offshore drilling:

Sea birds coated in oil and unable to fly;
Seas turtles poisoned by toxic waters; and

48

Wildlife habitat fouled by oil.
And it could be much worse in the Chukchi.

Thirty years after the Exxon Valdez disaster, there is still no
effective, proven technology to clean up oil spills in broken sea ice
conditions in Arctic waters, such as those found in the Chukchi
Sea. a problem that could doom rare Arctic whales, threatened
polar bears and other wildlife to extinction and destroy Inupiat
communities if drilling proceeds.

Help stop the next offshore drilling disaster! Tell President Obama
to pull the illegal leases sold in the Chukchi Sea and halt seismic
testing in the Arctic this summer.

The courts and scientists have all said that more information is
needed about the Arctic environment before we even consider
drilling in its ice covered seas that are cloaked in darkness most of
the year.

The recent court decision provides the perfect opportunity for the
Obama administration to take that time and really make sure we
can protect the environment – before we jeopardize this fragile
place, its wildlife and the people who count on it for the survival of
their communities.

To avoid another catastrophic offshore drilling disaster like the one
now threatening the communities and wildlife of the Gulf of
Mexico – and another Exxon Valdez-like oil spill – we need a
responsible approach to protecting the Chukchi.

The Valdez spill decimated fisheries and continues to impact local
wildlife and Alaskan communities to this day. And the Gulf oil
disaster has killed thousands of animals and will impact the
region's fragile ecosystems for years to come.

Alaska native communities, polar bears and the rest of America
deserve better. Please send President Obama a message right now.

Section - # 11 – Solutions:

No sense discussing the situation of Global Warming and Climate Control if you do not come up with ideas for solutions, so here we go:

Solar Panels:
Solar Panels are one solution, but do have cost and lack of sunlight for a good part of the day as a negative.

Wind Turbines:
Wind Turbines are a solution, but do have cost and lack of wind for a good part of the day as a negative.

Hydroelectric:
Hydroelectric is a solution, but dams block water flow and eventually the sediment builds up behind the dams and makes each useless.

Nuclear Power:
Nuclear power is a solution, but the waste product is a problem in that where and how do you dispose of a material that can remain radioactive for decades.

Biofuels:
Biofuels are a solution, but using plants for creating energy tends to remove plants for other things including food for the masses.

Geothermal:
Geothermal is a solution, and this author worked at the Middletown, California Geothermal Site for a period, but realize

that it is limited to places were we can easily get to the magma or heat sources.

Oil and Gasoline:
Oil and Gasoline is a solution, but we know that it pollutes the earth and has a tendency to destroy and kill.

Natural Gas:
Natural gas is a solution, but we know that Fracking the method of securing natural gas can cause earthquakes, and we also know that natural gas is explosive and can cause massive destruction if not properly handled.

Natural Gas is available to the USA, but the pipelines that supply it are very old in most of the country, and non-existent in the rural and farming areas, therefore, it will take trillions to 'plumb' in these areas.

The existing old pipelines in our cities and towns are estimated to be about 100 plus years old, and are failing as seen in California and Pennsylvania, and the Natural Gas industry states it will take the better part of 40 years to dig up and replace these now faulty lines. Not a good solution, but if we pipe the Natural Gas to power plants only, then we have a good situation where we can protect the limited numbers of pipelines, and generate electric that is wired to nearly 100% of America, and will be used for EVs and HS Rail.

The Fracking Wells producing Natural Gas are crossing the Delaware and destroying the Poconos, the Catskills, the Hudson Valley, etc., and the contaminated water from the Fracking is causing Utility Bills to rise due to NEW Water Plants having to be built that can detect the poisons and remove each.

Coal:
Coal is a solution, but we know that it can be as destructive as crude oil and is a killer.

So what is the answer? Hot Fusion, Cold Fusion, in space sunlight collectors, going back to burning wood, or?

Here are a few ideas on energy conservation:

Upside Down Energy Market:

The utility market is Upside down. The more you save and cut your use, the more per unit of electricity, gas, and fuel oil will cost you. Big Business, which uses tons of energy gets the best rates, while the POOR slob that is sweating in the summer and freezing in the winter in an attempt to lower his or her cost is paying TOP DOLLAR per UNIT. Also, if one wants to use GREEN energy, he or she will PAY MORE for it, than if they used DIRTY energy. We need to REVERSE this practice, and make it economical to save energy, and expensive to use it.

2:30 PM Jul 30, 2010

Saving Energy at Home:

We need a WPA or CCC organization to go to the homes of every American and advise on how a person can save energy, and if necessary help the person do the right thing, which is insulating, installing storm doors and windows, caulking, maintaining the water heater and furnace, maintaining the vehicles, setting up the thermostats, buying energy star appliances, adding individual room heaters and air conditioning, adding auto on-off lighting, changing incandescent bulbs to high output low energy use bulbs, and so on. We in the USA can save over 40% of our fuel if EVERYONE does these things.

Jet Fuel from Plants:

Our chemist and Green Energy companies are working to find new plant matter that can be used for fuels, and they have found that food waste in our garbage can be converted for use as a jet fuel. Here are some of the plants that we can use for making jet fuel.

Biofuels, Camelina seed oil -- homegrown in Montana

50-50 blend of jet fuel and oil from Jatropha seeds, grown in Malawi, Mozambique and Tanzania.

Boeing 747 left London -- without passengers, just in case -- while one engine ran on 20 percent oil from coconuts and Babassu nuts.

Can fuel made from weeds, nuts and pond scum keep people and freight aloft while helping keep global warming down? Many scientists seem to think so.

Tell Congress: End Oil Subsidies:

Big Oil doesn't need any more of our money.

There is no excuse for providing billions of dollars in tax credits and subsidies to cash-rich oil companies. I urge you to end all subsidies for oil companies in all future Federal budgets.

REF: https://www.fuelfreedom.org/oil-company-subsidies/

Push for Electric Surface Transportation:

The more we invest in EV (Electric Vehicles) the more we will save on transportation, and the more we shall cut pollution that is causing Global Warming, Climate Change, and massive illness across our planet.

Have you been to Denver Airport or Dallas Airport, both have a version of the 'People Mover' a horizontal elevator (Remote controlled train) that is efficient and non-polluting? We should be installing these between our cities, our town, and our residential to commercial areas.

See this author's Amazon.com/Kindle book on 'High Speed Rail: What you Should Know!'

Vehicle Use Should be Eliminated:

Our cities and towns are starting to take a look at Portland, OR, and it is looking good for America. The cities, like New York and Philadelphia are raising parking fees, parking meter fees, and are cutting down lanes so that there are people and bike lanes, instead of two or more auto lanes. This move is to 'gently' push people into moving closer to their jobs, closer to the city centers, and to get people out of the dirty air-polluting cars and into clean electric buses and light rail systems

Subterranean Construction:

Building underground has the advantage of a constant temperature and thus, constant heating and cooling that will be minimum. This form of building for homes and commercial building allows for surface land use of a green area, or pool, or parking, or recreational sports on small lots.

Grass Roofs:

Building 'Green' can be done by covering the building with dirt and grass that will hold the building's temperature inward, and thus cut the heating or cooling loads and cost. Surrounding your building with trees and deciduous trees will help with both heating and cooling during the seasons.

Water Pond Roofs:

Water can be used on the roof of a residential or commercial building to act like a heating and air conditioning system. The building can be cooled during the day by leaving the cover open at night and closed during the day, and heated by leaving the cover open during the day and closed at night.

Flowing Water Electric Generation:

We have in many cities and towns water supplies that come from water towers or through large pipes from mountain lakes. This water is constantly flowing and has pressure that can be used to turn turbines that generate electricity. The water supply from the Catskill Mountains to New York City is one such place where this would work well.

Overnight Water Storage, and Daytime Electricity Generation:

Water can be pumped to a mountaintop reservoir at night and then released to turn generators during the day when peak electricity consumption is in demand.

Road Surface Generators:

Major highways have a near constant flow of vehicles and these vehicles can be used to generate electricity by installing rollers along the roadway. The rollers would turn when a vehicle drive over each and each would turn a generator.

Hydrogen Generation:

This is under development and is a very 'Green' fuel as it burns clean with Oxygen and produces H2O or Water as its exhaust byproduct. Hydrogen can be extracted from our air and water, but is currently too expensive for most uses. The generators are also large and heavy and thus not currently suitable for vehicles, but can be for homes and businesses.

Railroad Track Edge Generators:

The idea is to have copper wire coils along the side of the railroad tracks and powerful magnets on the side of the moving trains. As the train passes the magnets it will generate current in the coils, and this electricity can be collected. This would work like a linear motor winding.

Burnt or Processed Garbage:

Some cities like Philadelphia are now using garbage for generating electric power. Others are attempting to use biodegradable garbage for jet fuel production.

Bovine Exhaust:

Farmers are finding that the Methane (cow gas) is lighter than air and thus rises. The farmers are adding collectors to the ceilings of their barns and collecting this flammable gas and using it to boil water that is used to turn steam engines and generators that produce electricity for the farm.

Gray Water Systems:

Gray water is the water you used for taking a bath, washing clothes, washing dishes, etc., it is not the water that is flushed down your toilet. Very few home have a method of collecting Gray Water, but we should as it reduces our need and wasting of fresh water (Which is becoming a commodity in many areas), and can be used for water outdoor flowers, edible plants, and lawns. All new homes should be equipped with dual waste systems, one for toilet water and one for Gray Water.

Education:

All high schools should be teaching a mandatory course in energy conservation and Global Warming, with emphasis on how we can build, eat, and live to protect the environment. Reading this book and researching the REF links (Website names provided) would be an excellent start at any age.

Index:

Cover Picture:

Ing-2920.jpg. Monterey Aquarium, California USA.

If you go to the Monterey Aquarium website you will find a listing of fish and seafood that are in jeopardy of being distinct, and a list of fish and seafood that are contaminated with industrial poisons and should be on your list to not eat. And a list of fish and seafood that is safe to consume.

https://www.montereybayaquarium.org/conservation-and-science/our-programs/seafood-watch

https://www.seafoodwatch.org

https://futureoftheocean.wordpress.com

https://futureoftheocean.wordpress.com/2019/08/13/california-action-alert-help-us-turn-the-tide-against-ocean-plastic-pollution

Other Environmental Books by the Author:

'High Speed Rail: What you Should Know!'
'Environment Tours in the U.S.A.'
'Border Security Solved with High Speed Rail: Generating Jobs using Existing Solutions'